G000164465

Traditional Ninja Weapons

By Charles Daniel

ISBN: 0-86568-075-2
Library of Congress Catalog Card Number: 85-52270

Unique Publications, Inc.
4201 Vanowen Place
Burbank, CA 91505

www.cfwenterprises.com

DEDICATION

For Martina, for always showing me that the sword that gives life is infinitely stronger and more courageous than the sword that takes life.

The author would like to thank Dr. Masaaki Hatsumi, Shidoshi Fumio Manaka, and Shidoshi Taro Yoshikawa for their teachings. Also, the author would like to express his thanks to Hank Reinhart and Konstantis Drevenis for reviewing this book and making suggestions about its presentation and format.

ABOUT THE AUTHOR

Charles Daniel is a yondan in bujinkan ninpo taijutsu, and received his instructor's menkyo directly from Dr. Masaaki Hatsumi. A prolific martial arts writer, Daniel has written articles for all the major American martial arts magazines. His first book, *Ninjutsu Nahkampf*, was written in German and published in Germany in 1984. He has also written on the historical development of Western martial arts weapons and techniques and is currently working on a book covering the development and use of the sword in Europe. Daniel is a graduate of The Georgia Institute of Technology.

About the Assistants

Ken Brooks is a sandan in bujinkan ninpo taijutsu and a former fencing champion. After earning a master's degree from The Georgia Institute of Technology, he began work as a senior industrial engineering consultant.

Konstantis Drevenis is a shodan in bujinkan ninpo taijutsu and a former judo champion. A graduate student at The Georgia Institute of Technology, Drevenis is a talented musician with a special interest in traditional Greek music.

TABLE OF CONTENTS

INTRODUCTION

The rise of "ninjamania" in the last few years has led to an incredible assortment of books, movies, fashions and even toys. While it is true that such an explosion of popularity has been a great financial windfall for many, all too often exploiters overlap the areas of fantasy and fact on purpose. This is not at all surprising, since much of the public fascination with ninjutsu is caused by popular authors' claims that the ninja were feared, all-powerful, and generally invincible. Of course, if this were true, Japanese history would be much different. There is no doubt that the ninja of old Japan were skilled martial artists, but this in no way means they were superhuman. Anyone claiming that the ninja had some unseen hand in the larger picture of world history should remember that secret political societies are always formed by groups who want political power—and don't have it.

For these reasons, this book is limited to the study of ninjutsu techniques as it applies to weapons. In this area, the imagination and intensity of the ninja reached a very high level. In fact, the methods and techniques in this book predated the rise of the samurai and their martial arts. To get a feeling for the importance of technique, one must take a mental step back into history.

In old Japan, families grouped together in clans much like old Scottish clans or American Indian tribes. In Japan, each clan developed its own special form of martial art. Thus, the Yagyu clan became famous for its swordsmen, the Kuroda clan was expert with the *jo* staff. While these two examples come from the Edo period, the practice of clan specialization is ancient, and well-suited to the Japanese way of thinking. So when one looks back into the past, one will see a Japan that is divided by both geographic and clan boundary lines. Within these lines, each clan trained and controlled its own army. Thus, if one clan member killed many enemy, he would teach his methods to the members of his own clan. This is probably one of the main reasons why some clans were particularly skilled in one area of the warrior arts. If one were trained in short spear techniques and was successful on the battlefield, then of course, one would impart the importance of short spear technique to the next generation. This would set a pattern for later generations. Obviously, if a clan relied on a certain technique (for example, short spear) to protect its domain, then an outsider's discovery of that technique could have dire results for the clan. For this reason, the early martial artists of Japan (often men from China) were valuable to the clans because of their deep knowledge of various techniques.

The word techniques has a slightly different meaning when one is talking about the methods of these early martial artists. Today, technique usually refers to a particular set of fixed motions such as a joint lock or sword block-cut combination. The earlier meaning of technique included various motions that had a common element such as center of motion or method of power generation. For example, in taijutsu, power is generated through proper knee motion, while in some Chinese kung-fu systems, power is generated through correct waist motion. The idea of approach is important because the time period here is the period of ca. A.D. 300-1490. The system of teaching known as *ryu* did not exist in the form it is known today until the year 1600.

The difference in these two approaches is simple: one is a martial art that developed in a period of war, while the other (after 1600) developed in a period of peace. Martial arts that evolve in periods of war are more general, using a wider variety of weapons. Thus, in ninjutsu, practitioners trained in swords, spears, bows and arrows, all of which were given a place of importance. In later, peacetime martial arts, weapons such as swords were given deep and special study while other weapons were ignored. Still later, there was a further split into armed and unarmed schools. The rise of such arts as *kendo*, *judo* and *karate-do* ended a long process of evolution begun with the enforced peace of the Edo period.

This short history of martial arts provides an essential lesson for today's practitioner. While there are many forms of martial arts in the world, many of these have nothing to do with fighting or self-defense. Many systems of karate and kung-fu are taught today to insure good tournament results. Other systems such as Thai boxing are concerned with special fighting skills for the area. In Japan, the ancient art of ninjutsu was concerned with espionage. The modern form of ninjutsu, called *ninpo,* is a form of self-discipline and self-knowledge which maintains its value as a self-protection system.

Weapons and the Martial Arts

In the sport forms of martial arts, the value of understanding ancient weapons is often overlooked. In fact, if one does not understand the principles of weapon usage, one is not a martial artist at all. Even the schools that stress unarmed fighting must understand how weapons work to protect themselves against them. Also, weapons can be great equalizers. Even someone with years of unarmed fighting training is at a natural disadvantage when faced with a weapon such as a knife or cane. Of course, training can improve one's odds, but there are never any guarantees.

The fact that weapons training can greatly improve unarmed ability is often overlooked. A shinai travels about three times as fast as a kick or punch. If one is skilled in fighting unarmed against weapons, then fighting unarmed against unarmed is not as big a problem. Training with weapons also makes one's timing and distance more accurate.

Note that these weapons are only important in terms of training. Few weapons such as swords and spears have any use besides teaching a high level of skill. In this day and age, the idea of carrying and using such weapons as ninja throwing stars or hidden knives is ludicrous, because there are far better modern weapons available. However, many of these modern weapons provide no backup in the event of failure, and the prudent martial artist will always take such weak points into account.

Which Weapon is the Best?

In today's martial arts culture, there is a ceaseless debate over weapons, weapon styles and the comparative strengths of this weapon over that. This debate even extends to the value of such weapons as hand guns and chemical weapons for self-defense. However, this book only considers hand-held or so-called "ninja" or "karate" weapons. This does not mean that guns, bows and arrows and blow guns are not ninja weapons, but covering all such projectile weapons is extremely confusing for the beginning martial artist and can often be difficult for the experienced teacher. One major reason for this entire situation is that people rarely consider the environment for which a particular weapon was designed.

Weapons can be divided into three general classes: 1) battlefield; 2) dueling; 3) surprise self-defense. These three classes each meet a different need and therefore, are designed and used differently.

Battlefield weapons are big, heavy and fairly simple. Examples are axes, spears, maces, giant swords and halberds. Since battlefield conditions rarely involved true "one-to-one" fighting, these weapons are designed to be used in big open spaces. These weapons are often made to exclude personal defense possibilities, since such defense was usually provided by armor. The range of these weapons is considerable, since these weapons are designed to attack in large formations. Under these conditions, it is often better to reach out to the enemy so that he can not break into your ranks and cause confusion. Such an infiltration could quickly spell the end of a large, well-organized fighting formation.

Dueling weapons are a special class of weapons, requiring special training and sometimes used under special rules. Such weapons as *katanas*, jo staffs and the West's rapier are all intended for dueling. Although these weapons could and did appear on the battlefield, they usually did so only as backups. The real purpose of these weapons is a test of skill between two men. In many cultures, such tests were considered a gentleman's right. These weapons require the most overall skill, because duels were fought under strict rules. A cheating duelist ran a very high risk of being killed by a second or even by the spectators. The mutual acceptance of high risk combined with a limited set of rules and weapons probably *explains* why many cultures credited duelists with high courage.

Surprise self-defense weapons include those weapons that the opponent learns about only too late. Popular weapons such as *nunchaku*, butterfly knives, walking sticks, weighted chains and many of today's "ninja" weapons fall into this class. With battlefield and dueling weapons, there is no real surprise for the opponent because he knows what he is fighting from the start. With surprise weapons he feels it first and sees it later. This, of course, seems like the perfect approach, except that almost all surprise weapons have the major drawback of having an extremely short range. Oddly enough, all unarmed fighting systems fall into this class of surprise weapons. Once an armed opponent discovers the trick of a surprise weapon, its value is considerably reduced. However, often the opponent never really has a chance, because by the time he knows the trick, there is nothing he can do about it.

This, of course, leads back the original question: "Which weapon is the best?" The only answer for this question is another question: "Best for what?" Each weapon system taught today began in a specific cultural environment. Just as each country has a different cultural and military history, so each weapon system will reflect that history. History also reveals an interesting and important aspect of each weapon. Across cultures (before firearms) the sword, spear, dagger and bow and arrow, appear with the most consistency. Herein lies an important key for the martial artist.

Martial Arts Misconceptions

There is a popular misconception in martial arts that hard training to build up strength and endurance is directly connected with actual skill. If this were true, then the best swordsmen would always be in their twenties. However, in old Japan, a man was considered to reach the peak of his skill sometime in his forties or fifties. Why the difference? The reasons are multifold. The basic problem lies with the modern confusion of sports and *bujutsu*. First of all, such sports as kendo or judo were never meant to have combat reality. The main purpose of these sports is to score points against an opponent. Obviously, when one involves oneself in such a contest, it is important to have endurance and strength for several matches. Also, as the sports developed, it became common for matches to become multi-point affairs. Thus, instead of a "one-hit-wins" match, kendo matches could be played on a three-point system. This type of scoring allows players to plan out strategies even if they lose the first touch. Such thinking is impossible in bujutsu. In the context of real fighting, the first cut or strike is usually the last one. Often, the combatants would strike at the same instant and kill each other. This occurrence of mutual killing should be taken into account for all weapon-related martial arts. It occurs often in duels between equally skilled, but not fully trained martial artists.

Knowing how to avoid this type of occurrence gives the older and more-experienced martial artist an edge. Younger, less-experienced martial artists have speed and strength, but lack judgment and rush in where they would have been wiser to wait. Also, the less experienced a person is, the less skillful is his training. The younger practitioner may be repeating training or over-training in a harmful fashion. Repeating training unnecessarily wastes time and over-training leads to injuries or worse, a general fatigue that lessens the martial artist's awareness for days.

The whole question of skillful training is extremely important to any martial artist. Without skillful training, one will just keep repeating earlier training without making any real progress. This is one of the main reasons why practitioners reach a level of skill and get stuck. Someone in this situation may change to a different martial art or quit training all together.

To train in a skillful manner, there should be an element of desperation. This means that the training is always aimed at pushing beyond one's limits. Thus, if one can fence skillfully against one opponent, another opponent should be added. If one becomes skilled against two opponents, a third should be added. Of course, one cannot keep raising the number of opponents—but one can continually raise the level of difficulty.

TRAINING, DISTANCE AND LEVERAGE

The study of any weapon and its use is always complicated. Not only must one take into account such fairly simple things as grip, stance and basic strikes, but one must tackle such concepts as proper training, distance, and environmental considerations. Only when one understands how these ideas apply to the weapon in hand is it possible to *begin* training. Such questions as how a particular weapon should be used against another can take years to understand and to apply in practice. One approach to this problem is the *kata* method.

The kata method of learning armed techniques has both its supporters and detractors. Before examining this argument, the two training methods should be reviewed.

The two-man kata method is often thought of as a strictly Eastern method of learning and teaching martial arts. This is not true. There are examples of Western sword kata from as long ago as 15th century Europe. This implies that the idea of a kata is universal.

A kata is a series of techniques or movements that are connected so as to create a training dance. These kata are taught in exact order and form, so that a kata learned today is an exact replica of the movement performed hundreds of years before. If this was all there was to kata training, then the detractors of the kata method would be secure in their assertion that kata are little more than dance training. However, nothing could be further from the truth. Training in a series of given techniques is only the basic level of the kata training method. Later, the student is expected to respond correctly during a kata when his teacher suddenly stops or even changes the order of techniques. The process of going from complete beginner to being able to respond correctly and exactly to a sudden unexpected change of events can take a long time. However, the process of training and the stages of development are *exactly* ordered. This makes kata good for training a large number of men. Because of this, kata training was associated with such samurai schools as the Shinto and Yagyu.

The opponents of the kata method charge that kata are empty exercises which fail to take into consideration the conditions of real fighting. Their preferred training methods stress sparring and sporting contests, such as foil fencing or kendo. This is the modernized approach to martial arts. The benefit of this approach is that it develops simple skills fairly quickly. The drawback is that it forces each individual to create his own style from scratch.

The two sides of this argument are right and wrong at the same time. For learning life-and-death "one-cut" fighting, the kata method is often far superior to the modern methods. In such situations, the straightforward techniques learned in kata training will crush the flashier sparring-centered method. Such a life-and-death frame of mind is not a part of sports or sparring methods like foil fencing or kendo. However, sports methods do have the advantage of allowing the practitioner to collect a great deal of "almost real" experience which leads to highly developed reflexes as applied to a particular sport weapon.

It's clear that kata people and sports-sparring people train with totally different ideas in mind. These methods have *no relationship whatsoever*. Kata training is kata and sports training is sport.

The training method of *bujin kan ninpo* falls outside both of these methods. The key to ninpo weapons training (and unarmed training) is the practitioner's realization of his own weak points (*suki*). Such realization leads naturally to the elimination of weak points. This is important to understand. In ninpo, the student does not train to be strong. Rather, he trains to eliminate his weak points. This is, of course, more an idea of not losing than of winning. However, in martial arts and weapons training, one's success or failure is determined by one's weak points. Strength is actually of little importance because weak points, if skillfully exploited, will end a conflict before a contest of brute strength ever takes place. This explains why more experienced martial artists can defeat younger and faster men. Both strength and speed are easily outclassed by higher technique.

When beginning this training, the student must learn to eliminate the most basic weak points by using proper stances and movement. Proper foot, hand, eye and body positioning must be studied. If the student's stance is weak, a skillful opponent can cut him down before he can make any move. Without proper stance (*kamae*), one simply cannot move out of the way of (much less counter) an attack. In weapons training, the process of learning stances is more difficult because of the added tension (physical and mental) caused by the weapon. For this reason, students learn the fundamentals of armed techniques through basic *tarhenjutsu* (unarmed) practice. For example, to learn sword technique, the student begins by studying basic unarmed stances and motions. Next the trainee learns basic cuts and thrusts. This weapon training should stress bold, free motions so the student can understand the weapon's reach and rhythm. Here rhythm means the motion of the sword as it moves to cut, and then returns to a ready position. So far, no weapon-against-weapon training has been done. At this point, the student knows basic movements of his own body and weapons.

The next stage includes basic unarmed avoidance drills. Here the trainee learns to avoid his opponent's techniques rather than overpower them. This type of drill teaches proper timing and minimum motion. By moving at the last possible instant, and then moving only the distance required, the student swordsman can give the opponent the impression that the attack will hit home. The student also positions himself so as to be able to cut down or disarm his opponent at his leisure.

Next, the student practices rolling, jumping, body drops and other escape methods with the weapon in hand. Along with these escaping moves, he practices weapon avoidance while holding the weapon. At this point, no parries or cuts are used. The student avoids the attack with simple, exact movements while holding the weapon still. This process of learning stance, weapon mechanics, unarmed motion and finally armed motion makes the student *very skilled* before he learns his first real technique. In addition, this process allows for an orderly discovery of the student's weak points. Also, by repeating these training exercises in a variety of conditions, the trainee sees how certain weak points show up in some circumstances and not in others. For example, when training on a wet and slippery surface, the student may discover breaks in balance that would not show up on a dry floor. Once again, the point is to learn what causes one to lose. This process also leads to an understanding of the opponent's weak points. While kata training methods are designed for only certain weapons, and sports training is even more limited, this weak-point approach gives the martial artist the ability to use whatever weapon is at hand.

KAMAE

The influence of stance on reach is evident in the following examples. The foot position most used in ninpo is shown. In this stance the front foot should always point directly ahead at the opponent. This makes it possible to extend one's reach with balance *without moving one's feet* (1 & 2). If one's front foot is turned inward (3) then such motions will cause a break in balance (4).

1

2

3

4

MOVEMENT

Besides the motion done without moving one's feet, there are two other basic motions which are much used. From the basic triangle stance (1), a simple single advance (2 & 3) or retreat (4 & 5) is used to make small adjustments in distance and position. For a larger adjustment, a cross-step (6 & 7) is used. This cross-step, which covers about twice the distance, is used for a simple advance or retreat.

Most spins or pivots are used after one of these two basic adjustment motions. In the following sword vs. hambo (8) example, a simple side-step by the hambo-armed defender allows him to evade the sword. Using an upward hit to the hands (9 & 10) the hambo man disarms (11 & 12) his opponent and then uses a pivot and drop to throw the opponent to the ground (13).

1

2

3

4

5

6

7

8

9

10

11

12

13

Distance

All martial artists agree that an understanding of distance is a crucial aspect of one's overall ability. Distance is the critical factor that determines when to attack. An attack initiated from too great a distance will be easily countered, while one started in close enough will strike down the opponent. Such tactics as fakes, misdirections and perception tricks are actually nothing more than attempts to get close enough to one's opponent to strike him down before he can counterattack. Learning what "close enough" means is one of the great obstacles facing the student martial artist.

Fundamental to the understanding of distance is the concept of reach. Reach is not how far one can cut or hit with one blow, but rather how far one can hit or strike *without moving one's feet.* This means that each individual has a particular reach that can be made longer by either correct body motion or by using a longer weapon. When the opponent is within one's own reach, it is possible to strike him before he has a chance to react. This principle applies to all situations—with weapons or unarmed—where the two opponents' weapons are *not touching.* Once the weapons are actually touching, then considerations such as timing and leverage become more important.

The concept of reach clarifies many of the environmental considerations of martial arts. By positioning himself so that the sun shines into his opponent's eyes, the martial artist eliminates his enemy's ability to see and judge reach. By making an enemy fight uphill, the practitioner reduces the enemy's reach while extending his own. In this situation, the man fighting downhill has the added advantage of gravity so his blow will naturally be more powerful. A tactic that is related to shortening an opponent's reach is positioning him so that he cannot maneuver. This will make his blows weak. For example, a swordsman standing in a narrow hallway with a low ceiling cannot cut powerfully because there's no space to swing his sword from an overhead position or from the side of his body. Along these same lines, a kicking specialist would be harmless when standing on a slippery surface or in water. It should be noted that armor and shields are really devices to weaken an enemy's weapon and therefore, indirectly, his reach. To summarize, environmental considerations are mostly an extension and refinement of the distance-reach concept.

The reach concept is also important in situations involving mixed weapons. This refers to sword against spear, or two swords against an axe, or any situation where two combatants are armed with weapons of different length, weight, and composition. In such situations, one must be able to determine the opponent's reach by evaluating his body structure and weapon type. Fortunately, this type of judgment can be developed through fairly basic training.

Even after extensive training, students should observe mixed weapons training. The man with the shorter weapon must keep in mind that against a skilled martial artist who is armed with a longer weapon, it will be almost impossible to hit home without "taking the opponent's blade." Of course, against an unskilled opponent, it is often possible to strike through a weak point in his guard or simply cut him down. As mentioned earlier, once weapons are in contact, leverage and timing become more important than distance. In such a situation, the man with the shorter weapon usually has a tremendous leverage advantage. This is why schools of short weapons such as *jutte*, short stick (*hambo*) and short sword have at various times enjoyed popularity. If the man with the shorter weapon can avoid or parry the first attack of the man with the longer weapon, then he can often close in and end the struggle with a short-range thrust, cut, or blow.

However, short weapon strategy does have its disadvantages. When armed with a shorter weapon and facing a longer weapon (e.g. short sword vs. spear), one finds it very difficult to make the first move and seize the initiative. Since one cannot seize the initiative, because his short weapon gives him a short reach, multiple attackers are very dangerous. Thus, to attack his long-weaponed opponent, he will have to step forward and try to engage his opponent's blade. During the time it takes to step forward, it is fairly easy for the man with the longer weapon to strike his opponent before he has a chance to close in. Therefore, when armed with a short weapon, one must wait for the opponent to make the first move or try to engage the opponent's weapon through his awkwardness or disrupted motion. These two principles show that short weapons are mainly defensive. While many cultures have evolved extensive systems of short-weapon fighting (particularly with the short sword), these systems were generally backups to longer and larger weapons, or used with the added advantage of surprise.

In contrast to the shorter weapons, the longer weapons such as spears, long swords and halberds are best used by making bold powerful cuts and thrusts. Longer weapons also make the idea of "beat attack" possible. Unlike the leverage techniques used with short weapons, the beat attack is a short smack to move the opponent's weapon aside or even disarm him. For example, a man armed with a short sword faces a man armed with a spear; both men point their weapons at their opponent. The spearman makes a sharp blow toward the short swordsman's blade and instantly thrusts home. Of course, the short swordsman could avoid this attack by controlling his distance. However, if he misjudges, he can be struck down before he has a chance to counter or escape.

Another theme in mixed weapon is that of "attacking during recovery." Whenever someone executes a technique such as a cut, there is a movement from stance to cut and then back to stance. Here stance means some form of guard position (*kamae*) from which one can attack or defend. Returning to a stance or "recovery" position does not necessarily mean returning to the original posture. Thus, when making a sword cut, one may take the sword downward from overhead and stop the sword with the point downward. This stopped position would be the recovery if the swordsman could easily attack and defend from the position. It is during this transition from attack to recovery that the long-weapon man is vulnerable. This weak point is often exploited by short-weapon specialists who nimbly avoid their opponent's large-weapon attack and then use a quick lunge and thrust or cut before their opponent can start a second move.

While large arching cuts are very powerful, they are difficult to perform without giving an opponent an opening. For this reason, weapon techniques that use very large motions were historically used on the battlefield by armored warriors. On the battlefield, large arching cuts with giant swords (*odachi*), battle axes (*oro*), or halberds (*bisentoh*), were used to bring down horses (and their riders), smash armor, break opponent's weapons and even cut down two or three enemies with one blow. When dressed in armor, the soldier did not have to be as exacting as when unarmored. Because of armor's ability to stop many cuts and thrusts, wide, heavy-edge blows were needed to bring down armored soldiers. In Japan, the influence of armor on weapons technique can still be seen in the sports of kendo and naginatado. In Europe, the invention of the rapier, which could pass through the joints in armor, and the later invention of the cannon, ended armor's importance.

Japanese kendo and naginatado reflect a form of armored sport fencing while European foil, epee and saber reflect unarmored sport fencing. Whereas kendo is still played on the pass, European sport fencing is not. Because of this, the European sport fencers of today use much smaller motions than their kendo counterparts. And it is the use of small motions both with one's weapon and body that make attacking during an opponent's recovery possible. For example, if a short swordsman is confronted with a staff man, then he can apply the following strategy: avoid the staff's blow by moving his body out of the way by just a few inches of the staff's reach. As the end of the staff passes by, the swordsman steps or jumps forward to cut whatever part of the staff man's body is nearest (usually the hand). In this example, the two weapons never touch. For his part, the staff man (spear man or long swordsman) could avoid this fate by *controlling the point of his weapon.* Control is critical for long weapons because once an opponent with a short weapon gets within (inside the point) of a long weapon, he can strike *without moving his feet.* The man with the long weapon must retreat (move his feet) to bring his weapon to bear. Thus, a critical principle for using a long weapon against a shorter weapon is always to keep the short weapon outside the *reach of the point.* That is, if the short-weaponed opponent can reach the point of the long weapon without moving his feet, then the man armed with the long weapon should either attack or step (or jump) away to increase the distance from his opponent.

DISTANCE

Few mixed-weapons situations require as great an understanding of distance as the classical match of sword against spear. In this example, the combatants both begin in gedan chudan nokamae. As the spearman lunges, the swordsman makes a slight adjustment with his back leg and parries the spear away from his body (1). This parry should be a small motion so that the spear misses by a small space. As the spearman withdraws his weapon, the swordsman, using leverage, follows. At the same time, he uses his sword to wind his opponent's weapon around (2). This winding motion is one method of controlling the spear's point. As he closes in, the swordsman throws his opponent's weapon up (3) and cuts (4). The cut may be followed by another cut.

1

2

3

4

Using the reach of a longer weapon to stop a shorter weapon's attack can be seen in the following example. Here, the naginata man begins in *tachinokamae* (5). He uses this kamae to tempt the swordsman into attacking. As the swordsman attacks, the naginata man uses a side-step to whip the blade of his weapon into the swordsman's advancing foot (6). If timed properly, this technique will unbalance the opponent (7) and make it easy to finish him (8).

5

6

7

8

Beat attacks also include methods to press an opponent's weapon, rather than to hit it out of the way. In this example, a naginata is used to pull the opponent's sword slightly, and then deliver a quick cut. From *chudanno-kamae* (9), the naginata man makes a small step, hooks his opponent's sword and pulls it outward and down (10). This is followed by a quick jump, and a snapping push-cut to the neck (11). The shaft of the naginata controls the sword blade. The naginata is then drawn back across the neck. This motion comes from under the knee (12). The blade is then flipped over and cuts under the arm (13) so as to throw the opponent.

9

10

11

12

Attacking during recovery is illustrated in the following example: From gedannokamae (14), the swordsman with the short sword avoids the spear thrust and at the same time uses his sword to contact the spear's shaft (15). As the spear is withdrawn, the swordsman drops his sword and uses his forearm to control the spear. At the same time, he jumps forward to deliver a thrust to the spearman's stomach (16). The jump not only quickly covers the distance between the opponents, but also adds to the power of the thrust.

14

15

Leverage

Once weapons touch, an understanding of distance must include such considerations as leverage and timing. Leverage and timing should be combined so as to take advantage of the motion of the opponent and his weapon.

Stated simply, leverage is using the strongest part of one's weapon against the weakest part of the opponent's weapon. For example, in *kenjuku*, if one contacts the opponent's sword with the back half of one's own blade on the front half of his blade, then one can easily control both weapons. The closer one's own hand comes to one's opponent's weapon (when weapons are already touching), then the more possible leverage. This is one important reason why short weapons such as short swords, jutte, and even *tessen* (iron fans) were sometimes preferred over long swords and spears. Because of their short lengths, these weapons give their users a leverage advantage over their opponent *if* the user avoids or parries the longer weapon's first attack. Often, men armed with short weapons would help their own odds by holding their short sword or jutte in one hand while throwing a shuriken, knife or blinding powder with the other. To apply leverage advantageously, one must step inside one's opponent's attack. To a great extent, the opponent's stance (*kamae*) will determine just how easy this stepping in will be.

If an opponent is unskilled, then it is often possible to step in through a hole in his guard (*suki*) and cut him down without bothering with his weapon. Such a weak point could open when the opponent moves his weapon back so as to hit harder. For example, when a swordsman who is standing close to his opponent raises his sword from middle position (*chudan*) to an upper position (*jodan*), the opponent can take advantage of this *suki* with a straight thrust to the throat or eye. The effectiveness of this move is attested to by the fact that it is illegal in sport kendo. For someone interested in learning real weapon technique, the lesson here is obvious. The position in which a weapon is held (*kamae*) determines what moves can be made without giving the opponent an opening. Distance from one's opponent is a controlling factor, in that the closer one comes to one's opponent, the fewer the alternatives available. Thus, at combat engagement distance, one must use attacks and defenses that allow one's weapon to travel in as simple a path as possible. Although this seems too easy an approach, it does eliminate many of the weak points that result from an over-reliance on fakes or tricks. When both opponents use this strategy, leverage becomes important because the weapons will touch, assuming a standoff doesn't take place. If two opponents find themselves locked in such a clinch, then a subtle conflict of timing can arise. While numerous techniques can be used in this situation, many can be dismissed because they lead to mutual slaying. That is, while many of these techniques do in fact cut down the opponent, they do so only by giving the opponent an opening through which he can counter-strike. To prevent this outcome, the ninpo student learns to develop sensitivity in his hands to tell exactly what his opponent is doing just by touching his weapon. This sensitivity allows one to judge the opponent's intention so that one can escape his technique while applying one's own.

Historically, the samurai's total disregard of his own life while cutting down his opponent (*ai-uchi*) is somewhat different from the attitude in ninpo, where one is taught to first avoid or escape one's opponent's attack. To slip away from the enemy at close quarters rather than standing still for a comparison of strength is just one example of this idea.

LEVERAGE

One example of using leverage and timing can be seen in the following sword-on-sword situation (1). One brute strength method to deal with this situation is to push down the opponent's sword so as to cut him (2). Unfortunately, this technique would more often than not lead to a mutual kill, as each man cuts the other (3). An easier and safer method is to glide under the opponent and cut to his hands (4). If one times this motion with the opponent's pushing, then his motion will add just that much more power to the cut.

2

3

When beginning training, the student is taught to time his technique so that his opponent misses him by only inches. In this example, the attacker begins in *jodannokamae* while the defender begins in *rainonokamae* (5). The attacker steps in and cuts downward. The defender waits to the last possible instant and then steps outside the cut (6). Then he pivots to cut the opponent's neck. Note the use of the sword (detail 7).

5

6

TRAINING IDEA

At first, avoidance drills should be performed while holding one's weapon still. This is important because a weapon can cause extra physical tension (particularly in the shoulders) and thus has an adverse effect on one's motion. As mentioned earlier, exercises with weapons are best done after some skill at unarmed weapon avoidance has been achieved. An example of such an exercise is as follows: the attacker (trainer) begins in jodannokame (1) and cuts downward. The trainee avoids the attack (2) while positioning himself so that he may countercut (3). Proper form, timing and distance are critical to this training.

1

2

KNIFE SKILLS

As a weapon and tool, the knife is easy to underestimate and overestimate.

It is often underestimated by unarmed fighting experts who feel they can easily defend themselves against it.

It is often overestimated when used against another weapon such as a club or short staff.

In most cases, the unarmed experts are right to think they can defend themselves against this weapon, since few people are skilled with a knife. Unlike the sword or staff, there are no schools that specialize in just knife technique. Because of this, most martial artists learn to defend themselves against unskilled attacks, such as overhand stabs or slashes to the body. These are dangerous, but also easy to counter. However, a skilled knife fighter will always control his distance when attacking or defending. He will also use the other weapons of his body such as his feet or unarmed hand. Thus, if his opponent grabs his knife hand, the skilled knife man will attack with a punch, or trip his opponent and not worry about his knife. This strategy makes the attack-wave concept critical: if the unarmed individual catches his attacker's knifehand or arm, he should instantly attack one, two or more weak points of the attacker's body. Such an attack wave will make the opponent quickly forget what he was planning to do. However, here too, there is a potential problem. A skilled opponent will often let the unarmed defender grab his knifehand just so the attacker can use another technique. The ninja call this idea *kyojutsu no kamae:* presenting the opponent with a false weak point so as to observe his counter, then attack the counter with a higher technique. Kyojutsu no kamae, combined with controlling one's distance, can be a problem for the unarmed fighter confronted with a knife. To solve this problem, one can:

1) Control the distance between oneself and one's opponent. One should try to stay far enough away so that he must take a step forward before making a cut.

2) Stay in continual motion. When faced with a knife, it is not a good idea to stop and set oneself in a solid fighting stance.

3) If unarmed, counterattack to parts of one's opponent's body that are not close to the knife. For example, kicks to the legs, groin or even knifearm can all be used.

4) Concentrate on not losing. One is under no obligation to prove oneself when faced with a weapon. Escaping from an attacker's danger zone should always be the foremost thought in the ninja's mind. If escape is impossible, then try to improvise some self-defense weapon. If that too proves impossible, then use unarmed fighting skills.

These ideas can easily be used regardless of a person's previous training or martial arts style.

It is equally easy to overestimate the knife as a weapon. A good knife is somewhat stronger than nunchaku or tonfa and somewhat weaker than a short staff (hambo) or weighted chain. Weapons such as nunchaku and tonfa are difficult and sometimes illegal to carry. Also, they are not designed to be weapons at all. They are ancient farming tools that became improvised weapons. Knives, along with weighted chains, maces, and clubs are designed to be weapons and often nothing more. The difference between weapon and non-weapon is important for the martial artist. Such weapons as nunchaku are best used as a surprise defense (if at all). The tonfa, nunchaku, and to a lesser degree, the knife, automatically lose much of their potential the instant the opponent becomes aware of them. Of course this could be said of all weapons, but it is better to hide small weapons until the instant they are needed.

One can practice using the knife as a self-defense weapon in a number of ways. Controlled sparring and drills with training knives made of wood, plastic or other hard material is an excellent method. This type of training should be non-competitive and the participant must train with everyone's improvement in mind. Hits or touches should only be called on oneself. Also, every touch one receives should be announced so that the training partner will know if his technique is working. Safety should be enforced at all times. The object of this training is to learn how *not* to be injured in a knife fight. If one is injured while training, then one is wasting one's time.

Here again some guidelines can be given for using a knife for self-defense:

1) Always attack whatever part of your opponent is closest. If he cuts at you, then control your distance and cut his hand. If he kicks at you, then control your distance and cut his leg or foot. Bear in mind that your opponent's clothing can influence your strategy. Heavy winter clothes and gloves can make powerful thrusts necessary. The nature of the clothing worn should

always be kept in mind when reading old knife fighting manuals. For example, old books from Germany show the fighters grabbing each other's knife blades. This tactic was possible in the 16th century because the fighters back then usually wore heavy steel gloves. The author of this old book did not bother to show the gloves in his illustration, because at the time everyone knew about steel gloves. What appear to be weak techniques may in reality be very powerful when both combatants are dressed in full armor or thick clothing.

2) Try to forget the sense of holding a weapon. Unlike many martial styles which teach the student to "think of the weapon as an extension of the body," the ninja is told to forget the weapon all together. This helps the student develop *zanshin* or total awareness.

3) Use continual motion to keep the proper distance from the opponent. Always be prepared to throw your knife if that will give you the chance you need to escape.

These guidelines are easy to follow, and at the same time are the safest way to deal with a knife. Historically, the knife and knife fighting have been considered extremely dangerous, because the opponents are so close together. However, by training with all the above guidelines, it is possible to safely protect oneself from this weapon.

KNIFE

The danger of simply grabbing an opponent's knife arm can be seen from the following two examples. In the case of a one-handed grab (1), a circular motion with the knife (2) will cut the opponent's arm and thus cause a release (3). In the case of a two-handed grab (4), the same motion can be used (5) and the free hand can be added for extra force. The use of proper body motion should be noted.

1

2

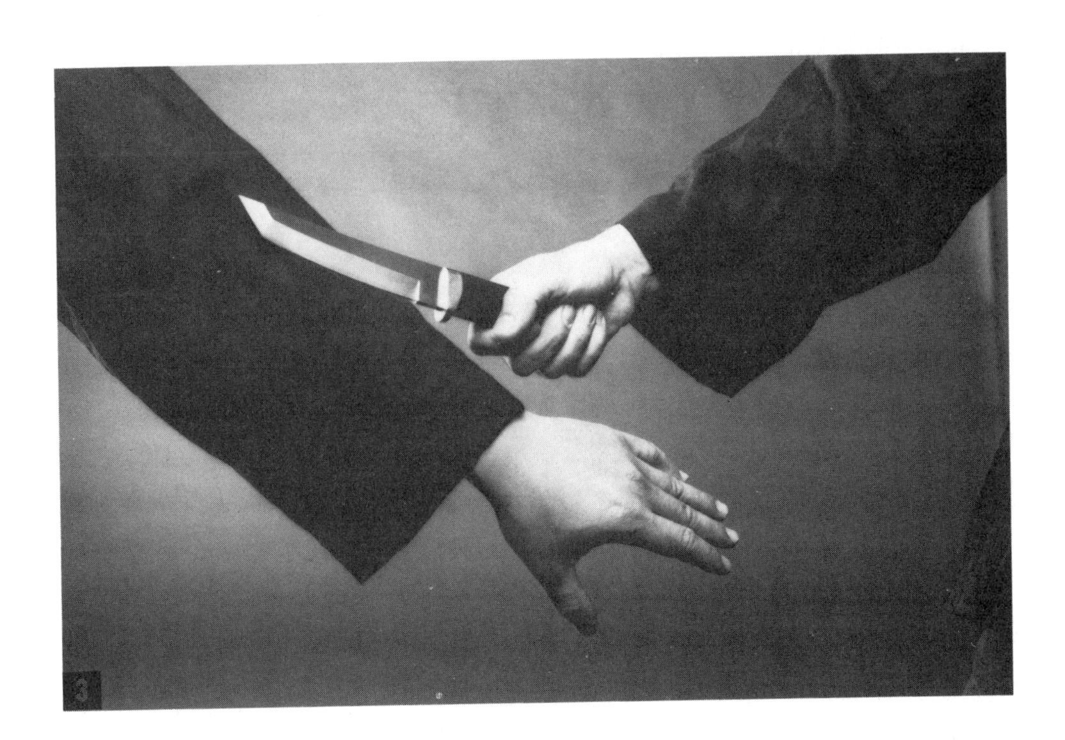
3

1

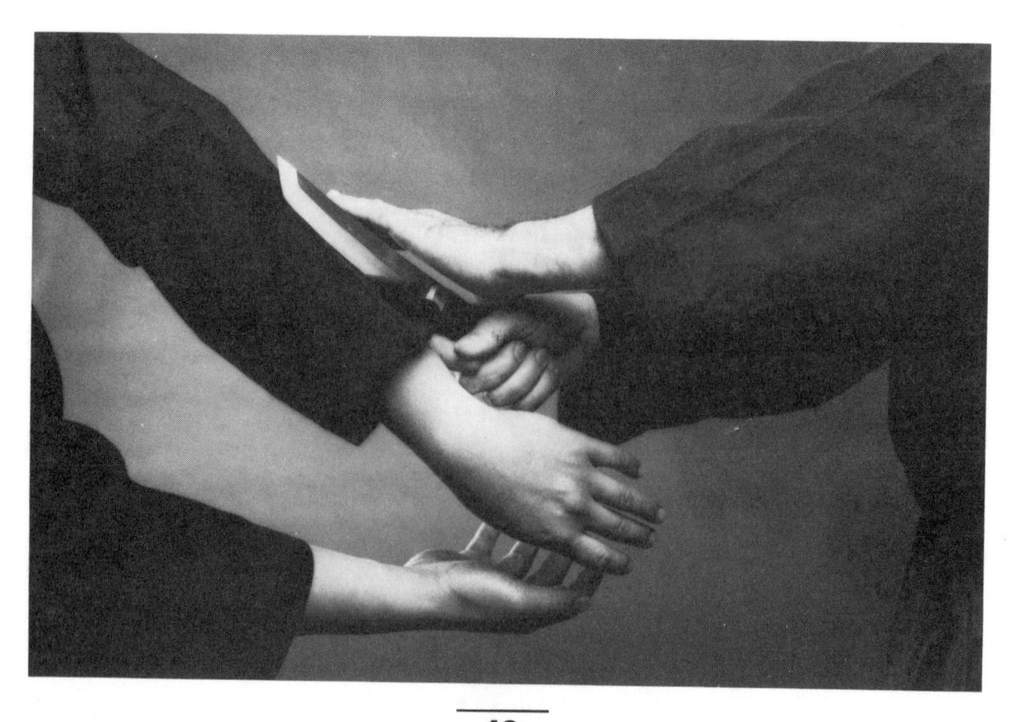

Here is an example of the attack-wave concept. The attacker begins with the knife held so as to make a powerful downward stab while the defender hides his knife behind his body. The attacker lunges and grabs the defender's front hand and raises his knife (6). As the attacker's blade descends, the defender cuts the opponent's arm (7) and then reaches across to cut the attacker's knife arm (8). Note that as he makes the second cut, the defender breaks the attacker's left arm.

6

7

Unquestionably, the safest way to deal with all attacks is with a powerful countercut to the attacker's knife hand or arm. If done properly, such a countercut will not only make the attacker drop his knife, but it could very easily incapacitate him. While making countercuts, the defender should at the same time move his body off the blade's line of attack.

Against a direct thrust, one can make body turns without moving one's feet (9). The defender spins to avoid a thrust at his stomach. While spinning, he uses his free hand to punch into his opponent's hip. Hooking and cutting the opponent's arm (10), the defender exerts pressure on the attacker's elbow to force him to the ground for a controlling or finishing technique (11).

9

10

11

In knife fighting, the tendency to jump straight back from an attack and thus give the attacker a second chance should be controlled. One method of doing this is by placing the trainee in a corner and making simple attacks at him. Because his backward retreat is cut off, he will have to learn how to move forward into an attack.
This drill should be done with the trainee armed and unarmed. At first, simple attacks should be used and later fakes added. This drill has special uses. If one can use a quick jump to escape, that is the best first choice. However, when this is impossible, one will realize that moving into or away from an attack is essentially the same thing.

Recently, a style of knife fighting using quick snapping slashes has become fashionable. These slashes are aimed at the hands, neck and face. Fakes are heavily relied on to cause the opponent to overreach, making it possible to cut him. This style is highly aggressive and can be more dangerous than any other method of attack. Example: The attacker makes a quick fake (1A & B) and snaps his hand back as the defender begins to countercut (2). As the defender's arm reaches full extension, the attacker makes a cut to the defender's hand (3). One of the weakest points of this system is its overuse of speed. During the actual fake, the knife snaps out and is snapped back. While the knife is moving backward—and even for a short time after it stops—the knife hand of the attacker is vulnerable because of the muscle tension required to make the sudden starts and stops this method requires. Also, fakes are always done out of range. It is the opponent's overreaction that effectively moves him into cutting range. The surest way to deal with this method is to be patient and remain just out of arm's reach by making slight adjustments as required. Thus, if the attacker makes a fake (4), a small adjustment (5) will place the defender out of range (6). This tactic will make it necessary for the attacker to make a fully committed attack by stepping in. This is fairly easy to counter. As one uses proper knee motion and controls one's distance, it is possible to cut the attacker during his recovery without moving one's feet.

2

5

6

STAFF WEAPONS

Of all the weapons in man's great arsenal, none could be more common than the ordinary staff. Actually little more than a polished stick, the staff was one of man's first weapons. Later, rocks and sticks were combined to create spears, axes and, still later, arrows. Bone, horns and animal teeth were also used in the making of crude Stone Age weapons. With the discovery of metals such as copper, bronze and iron, these crude weapons matured and became pikes, swords, armor and, more important, cannons. Yet in spite of these material improvements, the common staff still remains a useful weapon for a variety of reasons. On a mundane level, staff weapons are easy and inexpensive to make. It is also possible to strike easily with either end of a staff. On a more important level, a staff can be used to subdue an opponent without killing him. For this reason, staffs were a favorite weapon with priests, and are today favored by police and law enforcement agencies.

In ninpo, stick and staff techniques include a wide range of weapons from the tessen (iron fan) to extra long staffs (up to 12 feet). While such weapons as the tessen and jutte are made of metal, they are like a staff in that they use impact to stop the opponent. A middle-length staff is the hambo (three feet), or walking staff. According to Dr. Masaaki Hatsumi, the father of modern ninpo, the hambo is the original human weapon and the basis for all other weapons. Because of its size and length, the hambo is studied to learn principles that apply to both small and large weapons. According to Dr. Hatsumi, one should learn hambo techniques before learning sword techniques.

The power of stick techniques is easy to underestimate because sticks and staffs are not really weapons of war. However, in a one-to-one, or one-against-a-few combat situation, the stick has proven to be as good as, if not better than, many other hand-held weapons. Because of the hardness of certain woods such as white oak, a staff could shatter a sword blade with one blow. Also, a heavy club could be used to crush armor or knock down horses. Along with the usual array of staff weapons—hambo, jo, bo and club—the ninja also made use of such impact weapons as war mallets, hammers, *tetsubo* (heavy iron staff), and maces. All these weapons use impact rather than cutting to deliver a technique.

The advantages of heavy impact techniques are offset by two major disadvantages: their weight and the amount of force required for them to be effective. The disadvantage of weight is fairly easy to offset by training with heavy weapons. The disadvantage of having to hit the target very hard is not so easy to offset. Unlike bladed weapons which can inflict damage with a fairly light touch, impact weapons like staffs work best when hitting hardest. Thus, a lot of the subtle fencing skills of swordsmanship are not required. It is important to training to achieve the proper level of skill in body motion, timing and avoidance of the opponent's weapon. For this reason, many weapons such as bo staff and hambo are studied alongside some form of unarmed fighting.

The advantages of impact weapons are many and varied. Unlike swords which can be stopped by armor, an exact blow from an impact weapon can crush the best armor. Staff-type weapons often have the advantage of reach over bladed weapons such as swords. Staffs and clubs are also easy to improvise, hide, and replace when lost or broken. However, probably the biggest advantage of staff-type weapons is that *the techniques learned with stick and staff weapons will carry over to almost any other weapon.* A man skilled with a hambo and bo has a firm foundation for sword and spear technique.

The historical ninja's use of such unusual weapons as mallets, wooden posts and other "odd" impact weapons has a straightforward explanation. Historically, the ninja worked as information-gatherers and to a much lesser extent, as saboteurs and assassins. The image of a black-clad ninja sneaking over a castle wall to perform some task of devastation is mostly the product of popular imagination. However, as information-gatherers, the ninja would often adopt the role of worker or priest. Such roles naturally brought certain material implements into the ninja's hands. As a construction worker, the ninja could carry mallets, hammers, spades and other heavy tools used by that trade. As a priest or monk, such devices as ritual staffs, scrolls, bells and perhaps musical instruments such as flutes would be available. While it is true that some of these examples would not be ideal blunt weapons because of their light weight, others such as mallets and ritual staffs suffer no such shortcomings. It should be added that if a ninja ran a high risk of discovery, he could "improve" his tool-weapon by adding refinements such as chains or spring knives. These refinements were designed to increase the reach of the weapon to allow the ninja to reach the opponent while staying outside of sword range.

For the historical ninja, the understanding of sword range was critical. While it is true that the samurai were skilled with a variety of weapons, they used the sword the most frequently. Also, since much of the ninja's fighting was for self-defense rather than on the battlefield, he was sure to face the sword sooner or later. Because of this, many of the "surprises" used by the historical ninja added an extra three to four feet of reach to the impact weapon. This extra reach could nullify the reach of the sword.

The smaller impact weapons used by the ninja, such as jutte and tessen, were used in conjunction with taijutsu (unarmed method). In addition to being used in limited parries, these weapons could be used to strike nerve points and apply added pressure to joint locks. Of course, using such short weapons did place the ninja at a natural disadvantage. To offset these, there were a few tricks that could be used. For example, during the Tokugawa period, some ninja found employment as policemen. Because of the need to take prisoners alive, they favored the jutte. Under normal circumstances, a skilled swordsman could hold off several men armed with just a jutte. However, by moving in pairs or in three and throwing rocks, blinding powders, and cords, the ninja-police could subdue their target. It should be added here that the city-dwelling samurai of the Tokugawa era were for the most part mediocre martial artists. The level of skill required to disarm a skilled swordsman with such a short weapon is obtainable by only a few. So even in groups, the Tokugawa police had to be extremely careful if confronted by one of the rural swordsmen of that period.

The tessen or iron fan was essentially a backup weapon to be used when nothing else was available. Worn as much as a symbol of position as a weapon, the tessen could be used alone or in conjunction with a short sword. Because of its small size, many tessen techniques easily translate into self-defense methods in the modern world.

STAFF

The hambojutsu of ninpo serves as a basis for other weapons. The hambo can be held at the ends and used like a bo staff (1) or it can be held like a sword (2) to deliver blows. Basic thrust and hit motions can also be done with a jo (3, 4, 5). Thus, the hambo along with the jo incorporate practically all the motions that would be used with other weapons.

1

2

3

4

In addition to containing all the motions for major weapons, the hambo and jo can be used to grapple with the opponent and thus immobilize him. In this example, the hambo's reach advantage allows a knife attack to be overcome. From hiraichimonjinokamae (6), the defender side-steps the knife thrust and at the same time aims a blow to the attacker's knife hand (7). This blow should disarm the attacker. As a follow-up, the hambo man uses a technique called rolling up. He places the end of the hambo at the inside of the attacker's elbow (8) and uses a small circular motion to trap his opponent's arm (9).

7

8

Just as a hambo has a reach advantage over a knife, so too, a jo has a reach advantage over most swords. From hiraichimonjinokamae (10), the defender avoids a downward sword cut at his head and diverts a blow at the swordsman's hands (11). Next, a thrust is aimed at the opponent's throat (12). With the opponent stunned by these blows, it is easy to scoop up from underneath (13) and throw him. In both examples, the opponent is disarmed before grappling techniques are used.

10

11

12

13

Generally, if one can get the end of one's staff between the swordsman's hands, one will be in complete control of the situation. In this hambo against sword example (14), a direct staff thrust to the neck is used (15) against the swordsman's horizontal slash. The hambo is then used to throw and disarm the swordsman (16), then finish him (17 & 18).

14

15

16

17

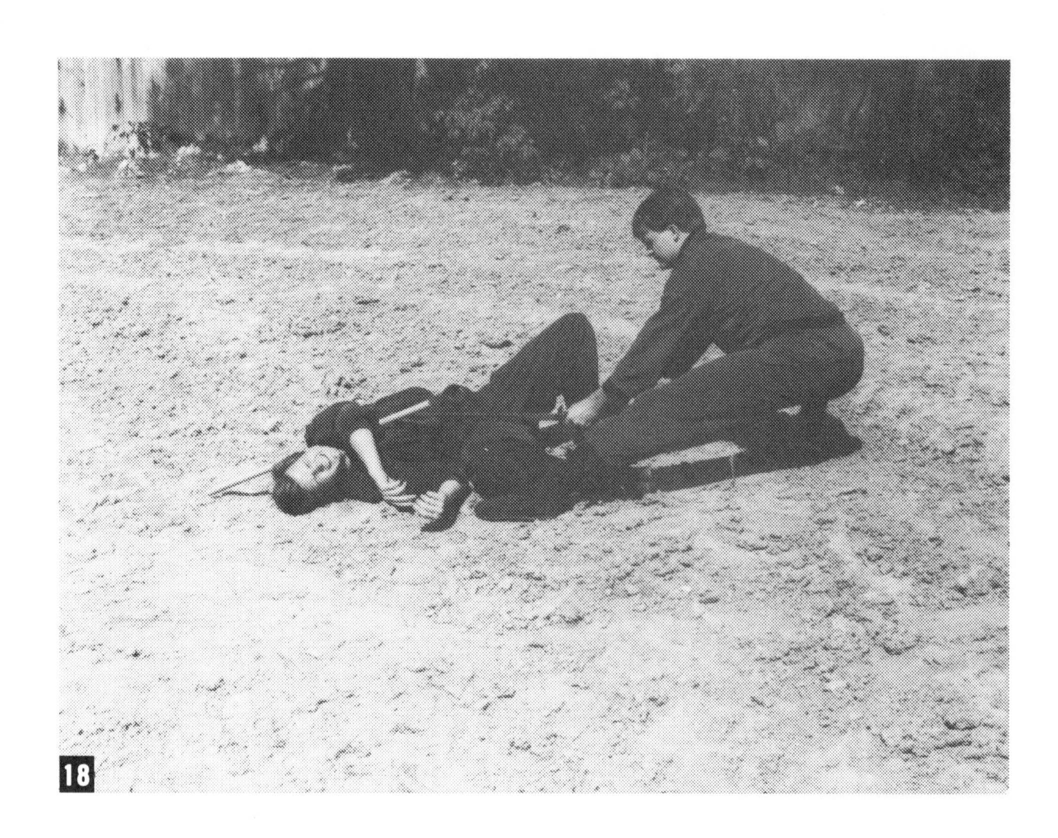
18

The reach advantage of the roshakubo over a sword makes it possible to retreat and strike at the same time. Beginning from tatsumakinokamae (19), the bo man retreats from a sword cut and at the same time spins his bo into the sword's blade (20). Then flipping his bo, he strikes his opponent's hands, thus disarming him (21). A quick thrust (22) ends the conflict (23). A trick that is often used in roshakubo against roshakubo is that of stepping on the opponent's weapon. Starting from jodannokamae (24), the attacker aims a sweeping blow at the defender's front leg. The defender steps slightly away from the attack and aims a blow to his opponent's staff (25). Then using a cross-step, he kicks downward on the attacker's staff (26) and brings the end of his staff up (27). He finishes with a sweeping blow to the neck (28).

19

20

21

22

23

24

25

26

27

28

When one is unarmed, or armed with a short weapon, and facing a long weapon such as a bo or spear, then the long-weaponed man's front hand is one's key target. In this tessen-against-roshakubo example (29), the tessen man begins in chudonnokamae while the bo man begins in tatsumakino-kamae. As the attacker makes a body strike, the defender steps straight in and hits the front hand (30). Catching the staff, he then traps the front hand (31). The defender finishes by throwing the opponent with his own staff (32 & 33).

29

30

31

32

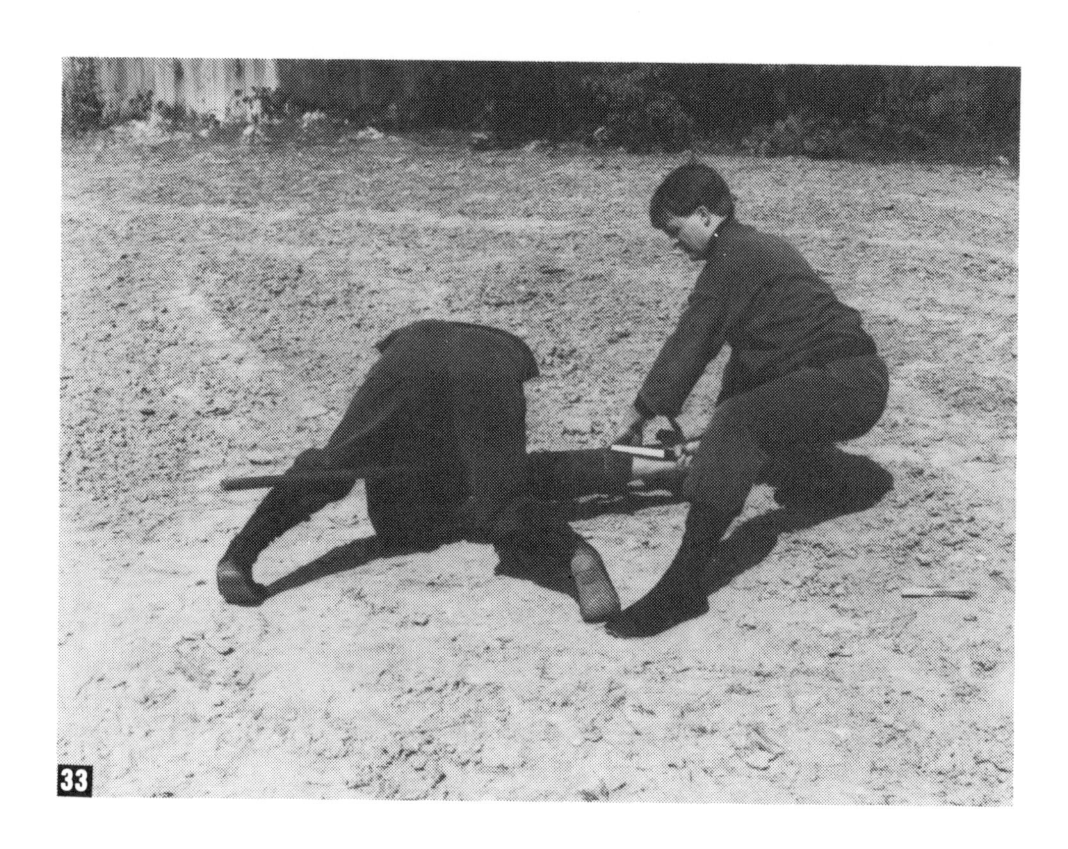

33

Another technique that is commonly used with jutte and tessen is combining those weapons' parrying abilities with kicking counters. Thus, the jutte is used to block an attack from any angle (34) while a kick is used to disarm an attacker (35) or knock him to the ground (36) where he can be controlled or finished (37 & 38). Twisting the jutte helps guide the sword back against the attacker.

34

35

36

37

Historically, the tessen or iron fan was often used by skilled swordsmen to defeat less-skilled opponents. This was particularly the case if one could not avoid a duel with a less-skilled swordsman, but had no desire to injure him. Tessen could also be worn and carried where other weapons such as swords could not. Because of its short length, a tessen could be drawn much faster than any sword. This made it a good weapon to counter *iai*-type attacks.

In the first example (39), the tessen is drawn from the belt and hits the opponent's sword hand (40). The tessen is then used to trap and disarm the swordsman (41). The tessen man then seizes the tassel with his free hand (42) and turns to choke the attacker (43). Another possibility from a ready position (44) is to step around the opponent (45 & 46) and pull his sword scabbard around (47 & 48); this is followed by hitting the opponent (49) to down him for a finishing blow (50).

39

40

41

42

43

44

45

46

47

48

49

50

SPEAR AND NAGINATA

In Japanese mythology, a spear was used to help create the nation's islands from the chaos of the primeval world. This myth suggests the importance and age of this archetype weapon. Undoubtedly, the spear in the form of a sharpened stick or a stick with a rock on the end was one of the very first weapons used by man or his prehuman ancestors. In fact, along with the bow and gun, the spear easily ranks as one of the most important hand-held weapons ever created.

With such a long history, it is hardly surprising to find an extensive practice of *sojutsu* or spear art in ninpo. Also, the ease of making this weapon made it particularly appropriate for the historical ninja.

In essence, the spear is designed to let one reach one's opponent while at the same time staying far enough away to remain untouched. This reach was probably first used to keep out of an animal's teeth or claws while still being able to kill one's prey. Later, someone discovered just how well this principle could be applied to fighting other men even if one were not particularly hungry. Still later, this little discovery led to ruling classes, slave classes, etc., and entire social units based on control by weapons and power. The discovery of copper, bronze, iron and steel brought the spear to its maturity. The spear would dominate man's battlefields for hundreds of years.

It is not known exactly when the spear was first used in Japan. However, over the centuries, the Japanese spear took on a unique appearance and application. There are two major reasons for the unique qualities of these spears. First, since the Japanese never really developed the shield, they could use their spears two-handed in an underhanded fashion (back hand pushing the spear from lower than shoulder height). In other parts of the world, the spear was used one-handed in an overhanded thrust so as to reach over the opponent's shield. A second reason Japanese spear use is unique is tied into the battle formations used. The Japanese formation was much looser than the formations used in Europe. This meant that each man had more freedom of movement to use his weapon. Also, since the Japanese saw war as a place to win glory through man-to-man fighting, their use of the spear was oriented toward the individual man instead of toward fighting in group formation.

According to Dr. Hatsumi, the first spears used by the ninja were knives tied onto long poles. As time went by, the ninja spear took a form all its own.

A short straight-bladed spear is used in basic training. This spear is approximately seven to eight feet in length and light in weight. In fact, it is much like a *roshaku* bo with a blade attached to one end. Often a metal end piece is attached to the other end of the shaft. This contrasts with the samurai spear, which is generally longer (nine feet to eighteen feet) and heavier, thus suited only for thrusting and hitting. According to Dr. Hatsumi, this combination of hitting and thrusting is necessary because a simple thrust is easy for an opponent to counter.

The similarity between the roshaku bo and the spear is another aspect of the technique overlap mentioned earlier. Technique overlapping is important in ninpo. That is, techniques learned with a bo apply to fighting with a spear and lessons learned with a hambo would apply to both. Since the spear is mostly a battlefield weapon, and battles produce many broken weapons, the hambo could be considered a broken spear.

With a steel tip and a long reach, the spear is extremely lethal. Unlike a bo or club which must land with a good amount of force to cause damage, a light tap from a spearhead can disable an opponent. In Japan, spears were generally as sharp as swords, and could be used to pierce armor. In training with a spear, a student must take several points into consideration. The reach advantage of a spear is obvious. However, this long reach has some disadvantages that may not be so apparent. For example, if one is standing in a narrow hallway, or worse yet, inside a boat, the spear's reach may actually get in the spearman's way. One solution is simply to break one's own spear under one's foot. Then one finds oneself armed with two half-spears (hambos) which work very well in cramped quarters.

Another interesting problem for a spearman is what to do if someone gets hold of one's spear. This is one reason the ninja used such a bewildering combination of cuts and thrusts in their spear method. It is common knowledge among most experienced weapons exponents that a thrust can be parried with very little force, while a cut is not so easy to deal with. This principle was often used by swordsmen to defeat spearmen. The swordsman would parry the spearman's thrust and then quickly grab the spear handle behind the spear head. This would immobilize the spear just long enough for the swordsman to deliver a cut or thrust. If the spear was extremely long, then the swordsman could resort to a number of spear-breaking techniques. The spearman had a number of ways to counter these tactics. The easiest method to deal with a grabbing technique was to use unarmed spear disarms in reverse. Thus, when one's spear was grabbed, one simply retreated a step or otherwise avoided his opponent's sword while at the same time using a disarm technique. This technique was often used with *kyojutsutenkan*, when one actually encouraged his opponent to grab his spear and then took advantage of a weak point shown by his opponent.

Despite all its advantages, the spear was not a weapon for everyone. Because of the amount of point control required to use the spear, the naginata was generally the weapon of first choice.

The ninja modified the classical Japanese naginata to fit their own needs. According to Dr. Hatsumi, the ninja's naginata was a little smaller than normal and the *tang* (blade shaft) inside the handle was longer. This is a "pawn's" weapon, used by men expecting to see the most fighting and thus with little chance of surviving. While the naginata did not require the same degree and type of dexterity as a spear, it did have many advantages when used in formation on a battlefield. Whereas a spear thrust would generally strike down only one enemy at a time, a wide arching cut from a naginata could cut down two or three men with one blow. Contrary to popular belief, the naginata was not used in an unending series of spinning cuts. The historical use of the naginata more closely resembled the use of a long-handled sword. Because of its long handle, the naginata could reach down to the opponent's legs or upward toward an enemy mounted on a horse. These actions could be done without exposing the weak points that would be revealed if the same were tried with a sword. The butt of the naginata's handle was used to parry and strike. The flat portion of the blade was also used.

Spearmen often use a series of thrusts and cuts in a flow of techniques. Each of the individual movements have a meaning and should be understood.

The most basic thrust used in ninpo is somewhat different from that used in most spear styles. In ninpo, the shaft of the spear does not slide through the hands. Thus, from chudannokamae (1), shift the body with the front knee while the hands guide the point into place (2). This is followed by a body shift to withdraw the body, hands and spear (3). Of course, the sliding thrust (4) is also used. However, by beginning with the non-sliding thrust, the student learns the body movement into and from an attack.

1

2

3

4

Because of its blade type and design, the Japanese spear could deliver powerful cuts. Here's an example of a basic cutting technique: from jodannokamae (5), the left hand releases the spear while the right hand rotates (6). The body is then shifted and the left hand grasps the shaft (7) to make the cut.

5

6

While both the spear and naginata are bladed, it's important to know the proper use of the other end of the shaft. In this example of naginata against sword (8), the end of the naginata blade is used to parry a sword cut to the head (9). Then a circular motion (10 & 11) is used to disarm the swordsman so he can be finished with a sweeping cut (12).

8

9

The combination of cut and thrust movements with a spear is seen in the following example. From chudannokamae (13), the spearman thrusts at a swordsman standing in chudannokamae. The swordsman parries with his blade (14) and then attempts to slide his sword down the handle of the spear. The spearman retreats (15) and cuts sideways to his opponent's arm (16).

13

14

15

16

Techniques involving catching the spear shaft are much used by swordsmen. From hiryunokamae (17), the swordsman parries a thrust at his face and slides one hand forward to catch the spear shaft (18) and cut the opponent (19).

17

18

When weapons reach a sufficient length, it is possible to counterattack and escape in the same motion. Here (20), a horizontal strike is countered with an "X" step and straight thrust (21).

20

21

KUSARI GAMA AND KYOKETSU SHOGE

The kusari gama and the kyoketsu shoge—two important ninja weapons—are really two forms of one idea. The idea is to combine a cutting tool with a long, weighted cord or chain. This combination can then be used as both a general tool and, if needed, a weapon. In old Japan, farmers used the classical weapon of a sickle combined with a weighted chain or weighted cord. Lumberjacks used an axe instead of the sickle. Battlefield warriors often used the rope and hooked dagger form of this weapon. The battlefield also gave rise to the *ogama* or giant chain and sickle. The chain of the ogama was extra long and heavy. These large weapons were fearsome and could be used against both mounted and foot soldiers. For purposes of simplicity, the name kusari gama will be used here, but what is said applies to all the above weapon forms.

The kusari gama is a specialist's weapon. Although used by samurai and commoner alike, this weapon never attracted the number of users that the more common sword, staff or spear did. There is a very good reason for this lack of popularity. These weapons take a very long time to master, and even then their effectiveness against a highly trained swordsman or spearman is uncertain. This is not to say that the kusari gama is not a dangerous weapon in the hands of an expert.

The history of the reknowned swordsman Shishido Baikin shows just how strong this weapon can be. During his lifetime Baikin fought and defeated close to 40 swordsmen in life-and-death struggles. However this is not the end of the story. Unfortunately for Baikin, he ended his career by fighting a duel with the famous Miyamoto Musashi. Musashi not only was a superlative swordsman but he had also received training in *yawara-ge* (a school of unarmed fighting, similar to jujutsu and *kakushi-jutsu* or the art of using small concealed weapons). When Shishido approached with his chain whirling, Musashi drew a dagger and threw it into Baikin's body. Thus, wounded, Baikin could present no defense to Musashi, who closed in to kill him with a sword.

The purpose of this example is not to glorify Musashi or the practice of dueling. However, there are some important lessons about the kusari gama contained in this story. First of all, the technique of spinning the cord or chain of these weapons is often just as dangerous to the user of the kusari gama as to his opponent. Whenever a long cord or chain is in motion, there is always the danger of becoming entangled in one's own weapon. This danger is made worse when one must dodge or move quickly in order to avoid an unexpected attack such as a thrown dagger. For this reason, it is probably better to hold the chain portion of the weapon still until the last instant. Then, as the opponent attacks, one should avoid the attack and at the same time throw the chain to catch him. It is also possible to hit with the weighted end of the chain. These blows can be very powerful, and when aimed at nerve points or bones, can easily disable the attacker.

Another important point to the story is Musashi's use of a throwing dagger. Musashi must have known the distance at which he could throw his dagger without endangering himself. Because Shishido was whirling his chain, Musashi could see how far the kusari gama could reach. Thus, it was easy for Musashi to stand just outside the chain's reach and throw his dagger. The lesson to be learned is that when one is using long chain or cord weapons, it is better to hide the length of that part of the weapon. This can be done by bunching the chain or cord and using a straight throwing motion. It is also possible to hold the weighted portion of the throwing cord behind the body and then strike by pivoting away from the attacker.

There is another set of ninja techniques which greatly increase the potential of the kusari gama. Small egg shells filled with explosives, poison liquids, or blinding powder increase the lethal potential of the chain portion of this weapon. Two other techniques along these lines should be mentioned. First, the ninja would sometimes bind a small poisonous snake to the chain or second, the chain was covered with a burning chemical. In either case, the opponent would be so busy dealing with the snake or burning chain that he could not present serious opposition to the ninja.

While techniques for entangling an attacker's legs do exist, these techniques are best used when one is armed with the larger and heavier kusari gama. The reason for this is straightforward. While it may be possible to trip an attacker armed with a sword, his sword will still be free. Closing with such an opponent can be dangerous. It is therefore much safer to either trap the opponent's weapon or aim strikes at his head, face, hands or ribs.

There is one last interesting and effective application for the kyoketsu shoge form of this weapon. This idea is easy to overlook, and yet when combined with surprise, it may be one of the strongest applications of this weapon. Simply make one quick twirl with the dagger end of the weapon and then throw the dagger at the attacker. This technique has several advantages. Because of the twirl, the dagger has much more power behind it than a thrown knife would have. Also, if the throw misses, it is easy to recover the dagger quickly so as to prevent an effective counter from the opponent.

Finally, the steel ring can be held in reserve so as to deliver strikes to the opponent. By using these techniques in combination with other strategies, one can ensure safe escape.

To summarize, the chief advantage of these weapons is their composite nature. Each section of the weapon should be mastered independently as well as together. The key to mastering these weapons is *maki*, or ensnaring the opponent's weapon with the chain or cord. To perform this action the kusari gama or kyoketsu shoge man will often avoid the attacker's weapon by just centimeters before throwing his chain or cord.

One should train with both weighted chains and weighted cords. The two materials have a very different speed and therefore timing will be different. Always take advantage of these weapons' reach. The key here is to stay just out of the opponent's reach and strike him from that distance. Always be aware that the opponent may try to rush in, so keep the sickle or dagger ready to parry such a sudden attack.

KUSARI GAMA

There are two basic methods for holding kusari gama weapons. The first is straight forward (1). The chain and sickle are thrown and then used in the right and left hand, respectively.

Another grip calls for the sickle to be held backward (2) and then shifted to the right hand after the throw. These same methods would also apply to the kyoketsu shoge.

2

In the following example, the length of the chain is hidden by holding it bunched in the hand (3). When the swordsman steps within range, the chain is thrown in his hands (4) and the sickle is used to knock the sword aside (5) and then cut (6).

3

4

The chain can be used to down an opponent, but care must be taken to avoid the opponent's downward blow. Against a roshakubo-armed opponent, the chain is used to pull the opponent's weapon past one's body (7A & B). The kusari gama wielder then fakes a downward cut (8) to draw the staff's blow. This is then parried (9), and the conflict is ended with a kick (10).

7

8

9

10

Kusari gama techniques can use many parts of the weapon in one flow of technique. Thus, against a spear (11) the kusari gama man throws his chain. This is parried with a backward sweeping motion (12) and a straight thrust, which are avoided by a side-step which lets the spearhead pass between the arm and the chain (13). A quick body twist traps the spear (14), and a sickle cut to the front hand (15) makes it possible to end the conflict (16).

11

12

13

14

15

16

A weapon which was used exclusively by the ninja is the kyoketsu shoge. This weapon is easy to conceal, has many uses and a particularly long reach (up to 18 feet). All in all, this is a particularly deadly weapon. The kyoketsu shoge was usually constructed with a long cord and heavy steel ring. While a cord does not have the bite of a chain, it is much faster and when combined with a steel ring has more dangerous impact. Whereas the chain of a kusari gama was used to entangle an opponent, the ring at the end of a kyoketsu shoge could kill him outright with a single blow. For this reason, the kyoketsu shoge could be used under almost any circumstances.

Here's an example of using the ring portion of this weapon: As the opponent swordsman steps in with a cut to the head (17), the shoge man steps straight in and blocks the cut at the base of the blade (18). He then uses his elbow to hit the opponent's hands up (19), and then uses the hook portion of the blade to finish the opponent (20).

17

18

19

NINJA SWORD

Of all the weapons in the art of ninpo, none has quite the special meaning of the sword. Dr. Hatsumi has referred to *taijutsu* (unarmed method) as *muto* or "no sword." Here, muto means swordsmanship without having a sword. This is really not surprising when one considers the level of skill required to fence "no sword" against a sword. Unlike bo, spear or chain, the sword leaves no margin of error because the entire length of the weapon is potentially lethal. Also, it is possible to grab a sword blade only in limited circumstances. Finally, the sword can be used to teach and refine timing and distance skills to a degree impossible with other weapons. Herein lies one of the secrets of the Japanese swordsman's success. According to modern measurements, a Japanese sword in the hands of an expert can move almost three times faster than the best unarmed punch. Obviously, if one can respond correctly to such an attack, then a punch or kick will appear to be traveling in slow motion. Thus, someone who trains with this weapon would be used to seeing what appeared to be extremely quick and powerful attacks and know how to respond accordingly. This type of training can also be used to discover one's weak points.

The actual techniques used in ninpo sword method vary according to the size and type of sword used and the opponent faced. The difference between an armored and unarmored opponent must be understood. For example, a light push-cut across the back of an opponent's hand is enough to end a conflict if he is not wearing any type of hand protection. However, often the samurai of old Japan did wear armor on their hands, so such a technique would have been useless. Some of the classical sword schools solved this problem by always aiming their cuts at areas that would be unprotected by armor. Such areas as the inside of the wrist, waist and neck would be vulnerable to cuts even if the unlucky opponent was wearing armor. In the same vein, the eyes, mouth and arm pit would be vulnerable to thrusts. This approach is practical on one level, but it could also lead to mutual slaying. While such a death might be honorable by samurai standards, it was not particularly desirable.

The ninpo approach to sword technique goes a long way toward eliminating this possibility by using proper body motion in conjunction with fencing skills. Parries are used to control an opponent's weapon rather than to block or misdirect the opponent. So even if the student ninja was to miss his opponent's blade entirely, he would still be safe because he would have moved his body out of the sword blade's path. In this respect, the ninja approach of "moving just the distance required, at the last possible instant" equals the approach of the finest samurai sword school. This should not be at all surprising because of the number of times the historical ninja and the samurai must have met on the battlefield. Certainly, both sides would have copied what they had seen work in actual application.

Actual training for sword technique is a multifaceted process. Just taking a sword out of its scabbard can be tricky business. Apparently simple skills such as cutting and putting the sword back in its scabbard can be dangerous, and true stories of men cutting or stabbing themselves with their own swords should always be remembered by those who would become swordsmen. Weapons, and swords in particular, should always be treated with respect. This respect will help reduce the chance of accidents.

The swords used by the historical ninja were as varied as their weapons. Giant swords of five feet in length were rare but not unknown. Shorter straight swords were also used. The weapon selected depended on what was available and the intended use. However, for the most part, the historical ninja seemed to have used smaller, shorter swords more often than any other type. This would be a logical choice because of the overall utility of such a sword. Unlike large battlefield swords, smaller short swords could be used in any environment.

What is a short-sword, and just what length of blade is correct for a given individual? Today's martial artists are generally larger and taller than 16th-century Japanese. This means that the majority of the weapons available through retail outlets are too small for the people buying them. If one is going to buy or make a weapon, that weapon should be appropriate to one's size, strength and height. Two examples will help clarify this problem.

First, the roshakubo of today is generally accepted to be the same length as that used by historical *budoka* 500 years ago. While this bo has retained a fairly constant length (approximately six feet), people's bodies have not. Thus, 500 years ago, Japanese men who were considerably less than six feet tall used a staff that was six feet long. Today, men considerably taller than six feet are still using a six-foot staff. The difficulties here are obvious. For a man of above-average height, the six-foot roshakubo would actually work better if used like a jojutsu (a staff less than five feet in length). Here is a better guideline for the proper length of these two staff weapons: the roshakubo should be up to one foot longer than a man is tall; the jo should be at least one foot shorter than a man is tall.

Second, the length of the sword is influenced by the change in man's size and height over the centuries. Here too, history has interesting and revealing lessons.

In Japan, the sword has retained its basic shape throughout that country's long history. However, changes in armor, battle tactics and simple fashions have influenced the sword's length, curvature and point shape. During wartime, the sword blade's average length varied from 30-48 inches. During the peaceful Edo period (17th-18th centuries), the sword's length became fixed at approximately 28 inches. It was during the Edo period that many of Japan's classical sword traditions came into existence. Thus, paradoxically, many of Japan's sword schools practice with swords and sword techniques that are actually different from those used in Japan's warring states period. This is not to denigrate these sword schools because of their unusual history. In fact, classical Japanese sword schools are some of the finest in the world. However, the approach used by classical schools is different than that used by the samurai in old Japan because of the difference in the length of the sword.

One of the universal principles of swordsmanship is that fencing skills only apply to swords within a certain range of length. At the lower end, blades become so short that one can parry against an opponent's arm instead of his blade. One can use swords with blades of less than two feet as one would use big knives. Of course even with such short blades, fencing skills will be necessary if one is facing an opponent armed with a longer weapon. When swords approach five feet, their weight and length will make many subtle moves impossible. As a general rule, short swords are more defensive and longer swords are more aggressive. Ideally, there should be some method of finding the right blade length for each individual. Fortunately, two such methods exist. In the first formula, a blade length should be ten times the width of a man's hand. In the second formula, a blade should be equal to the length of one arm plus the width of the shoulders. If one is above average in height, this could work out to a sword blade length of over 40 inches—considerably longer than the standard 28 inches mentioned earlier. In this case, the average Japanese sword falls under the category of short sword and can quite easily be used one-handed or double (one in each hand). In actual practice, the difference of two or three inches in blade length may not be important and it is possible to learn proper technique with a weapon of less than ideal length. However, weapon length and its relation to distance, timing and *body type* must be understood before one grasps the essence of sword technique.

Because of the different lengths of weapons, the historical ninja developed many methods for hiding their weapons' length from their opponents. For example, they would hide the weapon's length behind one's body or point the entire length of the weapon at the opponent's eyes. The first method is simple and self-explanatory. The second method takes advantage of the fact that if a straight weapon such as a sword or spear is pointing directly at one's eye, one cannot easily gauge the weapon's length.

Ninja also use tricks to momentarily blind an opponent to hide the weapon's length. For example, throwing blinding powder from one's sword scabbard is well-known. However, there is a subtlety in this technique that is often overlooked. Often an instant before the powder is thrown, the thrower will reflect sunlight off the blade of his sword, thus causing his opponent to blink. As his opponent blinks, the ninja throws the powder. The opponent opens his eyes just in time to catch a cloud of blinding powder. Another historical trick involved a lantern at night. If going out to face a particular enemy at night, the ninja could always carry a lantern with reflectors that made the light travel in a fixed beam. By shining the beam in the opponent's eyes and then suddenly putting out the light or even throwing the lantern, the ninja would momentarily blind his opponent and thus defeat him.

The fencing techniques of ninpo are varied, but center around the proper use of the last half-inch of the sword. That is, exact placement of the sword should be the rule for training. Actual techniques often combine parry and counter-cut or cutting two or three points in one movement. Also, the ninja uses the left or right hand to grab the opponent so as to deliver the final cut or to catch the opponent's weapon (e.g., spear).

An area of sword technique that seems to have been unique to the ninja was the creative use of the scabbard. Generally, when the samurai drew his sword to fight, he would also remove his scabbard and either hand it to an attendant or simply throw it down. Having a scabbard in one's belt while fighting is very dangerous because one can trip over it or one's opponent can grab it and use it as a weapon. So scabbards were generally pitched to be recovered later—if the owner survived. In all probability, tossing the scabbard before a fight became something of a fashion. The shocked horror of a samurai who watched his ninja or ninja-trained opponent draw his sword *and* scabbard can only be imagined. With a quick wrist snap the ninja would send a cloud of blinding powder or spring blade out of his scabbard. To take advantage of this tactic, the ninja wore his sword at his side much like the samurai. The popular image of the ninja with a sword strapped across his back is based on a misunderstanding. First, swords were worn across the back by both samurai and ninja as a method of transportation. This was particularly the case for *o-dachi*, which were from four to six feet in length. These swords were carried on the back to the battlefield. There the scabbard was discarded while the fighting took place. Fighting with a three- or four-foot scabbard strapped to one's back is out of the question. Some of the best methods for closing with an opponent,who is armed with a longer weapon (e.g., rolling),are impossible when one has three feet of wood strapped to one's back.

Several photographs of Dr. Hatsumi performing what appears to be a sword drawn from a back-wearing position have contributed to this misunderstanding. Actually he is doing a technique called *kaganoeto*, which is used when one cannot draw one's sword in normal *iai* fashion. This technique begins with the sword worn at the left side. Then the sword and scabbard are manipulated to the right side of the body so the practitioner can draw from right to left instead of from left to right, the normal method. Such seemingly small details will always separate the individuals trained in historical kobudo schools from those who are not.

SWORD

The sword's two basic cuts are the push cut (1 & 2) and the pull cut (3 & 4). Generally, pull cuts are much faster and stronger than push cuts. For this reason, pull cuts can be used anytime while push cuts are generally used after a parry, cut or beat attack. The danger of using a pushing cut as a first attack can be seen from the following series. From chudannokamae, the attacker attempts a push cut at the defender who stands in jodannokamae (5). Ignoring the attack, the defender cuts straight down at the attacker's hand (6). Because of blade angles and force directions, the defender's cut both parries and counterattacks.

4

5

The proper use of a push cut can be seen in the following series. The attacker cuts at the defender's head. The defender uses a slapping type parry while moving his body slightly off the line of attack (8 & 9). Then he rocks forward and pushes his blade across the opponent's throat (10). The defender's pushing cut throws the opponent back away from him.

7

8

9

Even when two opponents face one another with weapons that are exactly alike in terms of weight and length, they cannot fight on equal terms because their bodies aren't equal. The taller of the two would generally have a reach advantage and thus, it is possible to speak of natural advantages. The pride parents feel today at having a tall child may in some way reflect an obscure archetypal remembrance of the martial advantage of height thousands of years ago. Even today, sports such as wrestling, judo, and boxing are divided by weight class, and therefore by reach. Historically, men have often used their size advantage against opponents who had not reached the level of skill where their size and their opponent's size were irrelevant. For this reason, techniques were divided to help minimize size differences. The following technique is one such method.

This technique is known by many names but the most famous is Red Lewes Cut. It is best used by a smaller person against a larger person. Confronted with an opponent in chudannokamae, the defender adopts raikonokamae (11). Moving slightly to his right, the defender suddenly jumps and twists his body in the air and brings his blade down onto the back of his opponent's sword (12), causing him to drop his weapon (13). This blow is executed with the *back* of the sword. This technique should be practiced with wooden weapons, because it can often shatter the recipient's sword.

11

The method of holding a sword backhanded has been erroneously portrayed as an important ninpo technique. Actually, this method has very little to recommend it. Just as in knife technique, the backhanded grip is far too weak to be considered anything other than a "special case" technique. About all a backhanded sword grip (or knife grip) is good for is a downward stabbing motion. Of course, with a sword, the backhanded grip can be used to hide a blade. However, an experienced opponent would rarely fall for such a trick. Also, a downward stab is fairly simple to avoid. Attacks made with this grip are easy to deal with. From a backhanded grip, the attacker uses a horizontal slash (14 & 15). The defender simply retreats and uses a quick hit to the sword (16) and a thrust (17) to end the conflict.

15

16

It is often possible to cut the opponent in several places with one motion. The defender (18) steps from rensuinokamae to avoid an overhead cut, and countercuts to the opponent's hands and neck (19). He then push-cuts back across the same line (20), turns his blade over, and drops to cut the attacker's body (21).

18

19

20

21

The use of iai was well-known to the historical ninja. Because shorter swords can be drawn faster than longer swords, iai would naturally be favored by ninja armed with short swords. One of the keys to protecting oneself against iai type attacks (as well as modern quick-draw gun attacks) is to slam into the attacker. Thus, confronted with a sword attack (22), the defender slams his sword handle into the attacker's hand (23), grabs his head with his left hand, and draws his sword to cut to the neck (24).

22

23

24

CONCLUSION: NINPO TODAY

The purpose of this book is to give the readers an introduction to weapons techniques used by the ninja of old Japan. It is hoped that this short work will create an interest in the older schools of Japanese kobudo. However, the reader is to be cautioned. Licensed instructors are rare, and often hard to find. Often people who hold instructors' licenses do not have schools that are open to the public. However, there are schools run by men who are directly connected with Dr. Hatsumi's organization. These instructors' credentials are generally well-known and the interested reader should contact these men if they are interested in full-time active training.

As a closing note, any form of martial arts training should be considered as a way of defending oneself. Because of his knowledge through training, a true martial artist will abhor violence in any form. Just as animals in nature never fight unless their survival is at stake, so too a martial artist should only use his skills if there is absolutely no alternative. In the Western world, training in violence as a method of peace is still somewhat of a paradox. However, a man should be judged by how he behaves in the everyday world, regardless of his stated philosophy. No matter what style or school one teaches, if one's students learn to appreciate peace and value life, then they are true martial artists.

Unique Publications Book List

Action Kubotan Keychain • Kubota • 1100
Advanced Balisong Manual, The • Imada • 5192
Advanced Iron Palm • Gray • 416
Aikido: Traditional and New Tomiki • Higashi • 319
American Freestyle Karate • Anderson • 303
Art of Stretching and Kicking, The • Lew • 206
Balisong Manual, The • Imada • 5191
Beginner's Tai Chi Chuan • Chu • 280
Beyond Kicking • Frenette• 421
Bruce Lee's One and Three Inch Power Punch • Demile • 502
Bruce Lee: The Biography • Clouse • 144
Bruce Lee: The Untold Story • 401
Chi Kung: Taoist Secrets of Fitness & Longevity • Yu • 240
Chinese Healing Arts • Berk • 222
Choy Li Fut • Wong • 217
Complete Black Belt Hyung W.T.F., The • Cho • 584
Complete Guide to Kung-Fu Fighting Styles • Hallander • 221
Complete Iron Palm • Gray • 415
Complete Martial Artist Vol. 1, The • Cho• 5101
Complete Martial Artist Vol. 2, The • Cho • 5102
Complete Master's Jumping Kick, The • Cho • 581
Complete Master's Kick, The • Cho • 580
Complete One and Three Step Sparring, The • Cho • 582
Complete Tae Geuk Hyung W.T.F., The • Cho • 583
Complete Tae Kwon Do Hyung Vol. 1, The • Cho • 530
Complete Tae Kwon Do Hyung Vol. 2, The • Cho • 531
Complete Tae Kwon Do Hyung Vol. 3, The • Cho • 532
Deadly Karate Blows • Adams • 312
Deceptive Hands of Wing Chun, The • Wong • 201
Dynamic Strength • Wong • 209
Dynamic Stretching and Kicking • Wallace • 405
Effective Techniques of Unarmed Combat • Hui • 130
Effortless Combat Throws • Cartmell • 261
Enter The Dragon Deluxe Collector's Set • Little• EDSP2
Essence of Aikido, The • Sosa• 320
Fatal Flute and Stick Form • Chan • 215
Fighting Weapons of Korean Martial Arts, The • Suh • 355
Fundamentals of Pa Kua Chang, Vol. 1 • Nam • 245
Fundamentals of Pa Kua Chang, Vol. 2 • Nam • 246
Gene LeBell's Grappling World • LeBell • 593
Hapkido: The Integrated Fighting Art • Spear • 360
Hsing-I • McNeil • 225
Internal Secrets of Tai Chi Chuan • Wong • 250
Jackie Chan: The Best of IKF • Little and Wong • 599
Jean Frenette's Complete Guide to Stretching • Frenette • 420
Jeet Kune Do: A to Z, Volume 1 • Kent • 407
Jeet Kune Do: Entering to Trapping to Grappling • Hartsell • 403
Jeet Kune Do: Its Concepts and Philosophies • Vunak • 410
Jeet Kune Do Kickboxing • Kent • 526
JKD Vol. 2: Counterattack, Grapp. & Reversals • Hartsell • 404
Jeet Kune Do Unlimited • Richardson • 440
Jo: The Japanese Short Staff • Zier • 310
Jun Fan Jeet Kune Do: The Textbook • Kent • 528
Kata and Kumite for Karate • Thompson • 558
Kendo: The Way and Sport of the Sword • Finn • 562
Kenjustu: The Art of Japanese Swordsmanship • Daniel • 32
Kokushi-ryu Jujutsu • Higashi • 322
Koryu Aikido • Higashi • 321
Kung-Fu: History, Philosophy, and Techniques • Chow • 103
Kung-Fu: The Endless Journey • Wong • 230
Kung-Fu: The Way of Life • Wong • 202
Making of Enter the Dragon, The • Clouse • 145
Man of Contrasts • Cho • 508
Martial Arts Around the World • Soet • 140
Mas Oyama: The Legend, The Legacy • Lorden • 317
Ninjutsu History and Tradition • Hatsumi • 105
Northern Sil Lum #7, Moi Fah • Lam • 213
Nunchaku: The Complete Guide • Shiroma • 121
Pangu Mystical Qigong • Wei • 242
Practical Chin Na • Yuan • 260
Science of Martial Arts Training, The • Staley • 445
Searching for the Way • Sutton • 180
Secret History of the Sword, The • Amberger • 150
Shaolin Chin Na • Yang • 207
Shaolin Fighting Theories and Concepts • Wong • 205
Shaolin Five Animals Kung-Fu • Wong • 218
Shaolin Long Fist Kung-Fu • Yang • 208
Study of Form Mind Boxing • Tang • 235
Tai Chi for Two • Crompton • 568
Tai Chi Sensing Hands • Olson • 287
Tai Chi Thirteen Sword • Olson • 285
Tai Chi Training in China • Thomas • 567
Taijutsu: Ninja Art of Unarmed Combat • Daniel • 125
Tang Soo Do • Lee • 585
Tao of Health and Fitness, The • Miao • 270
Traditional Ninja Weapons • Daniel • 108
Training and Fighting Skills • Urquidez • 402
Tomiki Aikido: Randori & Koryu No Kata • Loi • 551
Total Quality Martial Arts • Hess. • 447
Ultimate Kick, The • Wallace • 406
Warrior Walking • Holzer • 155
Warrior Within • Brewerton • 450
Wing Chun Bil Jee • Cheung • 214
Wu Style of Tai Chi Chuan, The • Lee • 211
Xing Yi Nei Gong • Miller • 226
Yang Style Tai Chi Chuan • Yang • 210
Yuen Kay-San Wing Chun Kuen • Ritchie • 275-1

UNIQUE PUBLICATIONS
4201 W. Vanowen Place
Burbank, Calif. 91505

(800) 332-3330
www.cfwenterprises.com